CW00501119

A

HALF A STRING

Production

Based on Half a String's original theatre production
Breathe (2023).

www.halfastring.co.uk

Supported and funded by Arts Council England.

Self published in 2023 by Half a String in Canterbury, England.

ISBN 978-1-7392224-1-3

Written by Louisa Ashton
Story by Peter Morton
Illustrated by Caleb Simmons

Based on 'Breathe', a theatre production
made by Half a String

Walk to a woodland.
How still the trees. How quiet the earth…

But really, the trees are never still.
And the earth is never quiet.
The woods are as busy as a city…
… and as strange as the things we see in our dreams.

Now it's Autumn.   Leaves everywhere…   Mushrooms!
Lots of things for little mouths to eat.
Creeping through munchy crunchy forest floor,
worms and bugs and crawly things.

And far above, standing tall, slow sky-reaching giants…
The TREES.

Ancient trees are home to thousands of
different types of bug and plant and mushroom and moss.
And the trees shelter them, feed them, talk to them.
Trees are the LOW BEATING BOOM that lead the forest's rhythm.

Like OAK trees.

An oak tree can produce about
half a million acorns in its lifetime.
And yet...
Sometimes, only one of those
acorns will grow to be
 a new oak tree.

Just one.

Like this small soul. An ACORN.

How can something so small
grow so big and do so much?

Well... let's see.

A seed is alive whilst it waits to grow.

Some seeds can wait for up to a thousand years.

1 in half a million acorns will survive and grow to adulthood.

This acorn. Small. Smooth. Shiny.
It holds on tight to the mother oak, until…
A strong wind blows and it loses its grip!
It falls and spins and bounces on the
mulchy mess of the woodland floor.

So here's something special. An oak seedling!
Imagine. This little Seedling can
grow to be something so big!
Imagine. That this Seedling
might be that one in half
a million that grows
to be an oak.

Seedling wants to stand still and
look around the woodland floor,
but there's no time! No space!
It's Autumn, and it's busy.

It's hot down here.
Wet from rain.
The ground is damp!
Leaves and twigs and bits of bark.
Everything is turning soft.
Everything is crumbling into tiny
little bits to make soil
and humus [HYOO-MUS]
And humus makes the earth rich
and ready for growing things!

Seedling is pushed aside by a woodlouse.
Skiffffff skiffffff skuffle skiffffff.

Seedling hears something furry scuttle by.
Chatter chatter chit chit chatter.

Seedling feels the ground is alive with tiny things eating.
Num num munchy yummmmmmm.

Something is tapping the tree up in the tree tops!
Tic tic tactac tic tic tactic.

And the wind runs fast through the branches high above.
Shhhhh… booooooooooooom… shhhhhh…

Where is Seedling supposed to be?

Ancient oak trees can support up to 23,000 different types of wildlife.

Seedling feels the trees far above shiver. What's coming now?

SPLOOOOOOOPSH... a big blob of water falls on Seedling.

Here comes the rain!

It falls and hits the brown, red, grey, orange earth like a drum.

WHOMP THUMP THUD.

Something is moving through the humus.

WHOMP THUMP THUD.

Worms! On top of the soil, here to dance and coil.

They curl and stretch and roll and wriggle.

WHOMP THUMP THUD.

And from the humus, fungi [FUN-GEE] grow.

They race each other, growing fast.

Green Elf Cups! Tiny and bright.

Spindleshanks! Brown and bobbly.

WHOMP THUMP THUD.

Worms have 5 hearts?

Oak Jewel Beetles love eating oak leaves

Fungi is the whole organism. Mycelium are like the roots. Mushrooms are the fruit.

Seedling is lost. Seedling is dizzy.

Seedling ducks quickly as something whizzzzzzes past them.

Blue one moment, green the next. Shimmering. Fluttering.

An Oak Jewel Beetle!

Everyone is here!

A PARTY in the rain before winter comes!

But Seedling doesn't know this dance.

Where is Seedling supposed to be?

Suddenly, Seedling is whooshing up and up and up…
Caught inside a woodpecker's beak!
Acorns are a tasty autumn treat.

The higher they go, the further Seedling can see.
Tumbling lines of water, green land, grey rock, and trees…
Tall trees, wide trees, bendy trees, twisted trees.
Seedling holds a breath. The world is so much
bigger than they thought.

The bird's beak is getting tighter around Seedling.
Seedling doesn't want to be lunch, not today!
They wriggle and squirm. Squirm and wriggle.
Until…
Surprised and unsure, the woodpecker lets go!
Seedling is free and flying, falling and whirling…

Seedling lands as heavy as rain
with a THUD.
They look up to see their mother
tree. They haven't gone far!
But the ground is different here.
The rain has washed away the soil
to reveal large oak roots.
One mighty root creaks and
groans and looks at Seedling.

Who is this that comes?
Comes among us old oak roots?
Comes to join the humming
thrum of the earth?
A seedling? As winter comes?
Lucky to be alive!
Stay here and learn the
language of soil...

This is mycelium! [MY-SEE-LEE-UM]
Above the soil, we see the mushrooms...
but underneath the forest floor –
there's so much more.
From a mushroom's foot grows mycelium!
Hundreds and thousands of thin fibre-threads.
Like spider webs. Like strands of hair.
Like tiny roots, stretching out for
miles and miles.
Chattering and talking!
Messengers of the soil.
Music makers of the earth.

Mycelium is too small to see with the human eye!

Funghi are not plants, they are more like animals

Trees can warn other trees of danger using mycelium

Seedling listens and learns of mycelium.
How they spread, linking tree and grass and plant.
How everything in the earth reaches for each other.
How they listen without ears. And Seedling thinks
what other wonders are out there?
How can they be a part of it?

Mycelium can help healthy trees transmit nutrients to sick trees!

At 1500 years old, the oldest living thing on earth is a mycelium network

A spoonful of soil can have up to 60 miles of mycelium

Seedling touches the mycelium and
feels an old, old rhythm.
Thrrrruuuuummm…
A beat that feels like home.
A pulse that shakes them. Calls them.
Thrrrruuuuummm…
Seedling knows this thrum, this thrill, this hum.
The voice of their mother tree.
Thrrrruuuuummm…

Holding roots and threads,
Seedling pulls themselves up, up, up
to rough crinkled bark.

Clinging tight to the mother oak's trunk,
Seedling feels the air is changing.
This is something new.
A cold wind that bites Seedling's skin.
Winter has arrived.

Keep climbing, keep going!
Even though the winds are fast up here.
Winds for blowing small seedlings away.
Keep climbing, keep going.

And then they see it. They feel it.
An opening in the bark.
An old scar from a battle long ago.
The pull of the pulse is strongest here and
every atom of Seedling knows it.
Thrrrruuuuummm…
There is something wonderful and frightening
and familiar in this song.
Thrrrruuuuummm…
Trembling, breathless,
Seedling follows the hum within.

The tree's scar runs deep. Seedling pushes past layer
after layer, minuscule motorways running from
earth up to branch!
Thrrrruuuuummm…
Tiny tubes busy with tree business. Tubes carrying
water. Sweet, thick sap oozing back and forth.
Dark rings that tell us the age of the tree.
Thrrrruuuuummm…
Seedling climbs deeper… until everything they touch
hums and beats and sings.
Thrrrruuuuummm…

And here at the heart of the tree
Seedling finds a place to listen.

This is the tree's heartwood.
Old wise wood.
Seedling listens to the hum of their mother tree.
She tells Seedling how her toes tingle with mycelium
and how her hair flies high in the strongest of winds.
She tells Seedling how the earth turns slow and
steady. How colours shift from green
to orange to white.

She tells of how a seed must
learn to wait before they
can root and breathe.

Feeling braver than before, Seedling leaves
the heartwood.
They climb out to the oak bark, the heartwood
rhythm still beating in their chest. They are higher
than they thought!

Seedling shivers. Their limbs hurt with cold.
They see the turning change their mother tree spoke of.
Now, all is white and glistening.
But they are not alone…

Here she is.

The Guardian of the tree tops.
Made from the last leaves of the year!
She watches for Spring.
She watches the sleeping trees.
And in the depths of winter, they're sleeping deep.

But what's this she sees?

A Seedling in winter? Up here?
Lucky to be alive! Stay with
me, young one.
Stay and watch the wonder of
what winter can become!

Stay here, where the
winds CUT and thud and
SSSWWwwooooOOP with
icy breath.
Watch the land shudder.
Watch the world rest.
The dark is rising and the
nights are long.
No time for a seedling.
Not yet. But soon.

Soon it will be time for small
things to grow stronger than
they could ever imagine...

Trees crystalise their cells over winter to protect them from the cold.

During winter, trees store food and water in their roots.

Trees choose to lose their leaves, in the winter to help save precious energy.

Cold days become warm and bright.
The wait is over!
Look how spring wakes up!
How the leaf buds uncurl!

The Guardian smiles one last time
with all the warmth of spring.
Then, with a sigh, she becomes
the wind and the air and the sun,
falling to the ground to be
with the earth again.

It's hard, but Seedling is
beginning to understand.
How things grow from the
earth and return to it.
How change means the loss
of some things, and the
beginning of others.
All it took was the patience
of a tree and the strength
of an oak.

Now spring is finally here…
And Seedling feels different.
Changed. Refreshed. Awake!

leaves are actually red or purple! It's sunlight that turns them green

There are as many leaves on a tree as there are hairs on your head!

Listen. Feel that? Everything is moving,
everything is beating out its song.
Skiffffff skuffle.
Chatter chit chit.
Num num munchy yummmmmmm.
Tic tic tactac.
Seedling knows this music.
And now they know how BIG their beat
can be in this woodland full of sound…
Green hair blowing in the wind.
Feet rooted deep into the earth.
Shhhhh… booooooooooooooom… shhhhhh…

This is the time for a Seedling!
Grab a leaf! Glide through the sun!
WWWhhhooooooooSSSHHHHH.

Only one in half a million acorns grow!
Maybe our Seedling is one of those?

As Seedling glides from the tree tops, they can't believe…
Everyone survived the cold. Each plant and tree!
Each Oak Jewel Beetle and woodlouse and bird.
Green and gold and ochre and orange! Everything is here!

Seedling lands close to their mother tree.
They snuggle into the soil and then... they wait.

Slow. Slow. Roots grow down towards the dark.
Small green leaves stretch up towards the light.
And as the woodland turns gold to brown to
white to gold again,
Seedling becomes a sapling child.
And they keep growing.

Their stem will grow wide, thin twigs will grow strong.
Their roots will reach out to touch other music makers
of the woods, and soon they'll grow
their own little acorns.
One in half a million.

Our Seedling found their space.
Space to plant their feet!
A chance to breathe.

And all the woodland drums to the
song of a new oak tree.